Metric Conversion Chart

Symbol	When you know	Multiply By	To Find	Symbol
g	grams	0.035	ounce	oz
kg	kilograms	2.2	pounds	lb
t	tonnes (1000 kg)	1.1	tons (2000 lb)	

If it's mass or weight you're talking about...this chart will really knock you out!

by Ann Segan

Prentice-Hall Inc.
Englewood Cliffs, N.J.

for
♥ Leila and Sascha ♥

Copyright © 1979 by Ann Segan
All rights reserved. No part of this book may be
reproduced in any form or by any means, except for
the inclusion of brief quotations in a review,
without permission in writing from the publisher.
Printed in the United States of America J
Prentice-Hall International, Inc., London
Prentice-Hall of Australia, Pty. Ltd., North Sydney
Prentice-Hall of Canada, Ltd., Toronto
Prentice-Hall of India Private Ltd., New Delhi
Prentice-Hall of Japan, Inc., Tokyo
Prentice-Hall of Southeast Asia Pte. Ltd., Singapore
Whitehall Books Limited, Wellington, New Zealand
1 2 3 4 5 6 7 8 9 10

Library of Congress Cataloging in Publication Data

Segan, Ann, 1949–
  One meter Max.
    SUMMARY: Max Meter, Minnie Millimeter, Cedric Centimeter, and Katie Kilometer explain metric units of measurement.
  1. Metric system—Juvenile Literature. [1. Metric system] I. Title

QC92.5.S43   530'.8   79-15150
ISBN 0-13-636076-9

SEE HOW IT ALL BEGAN...

METRIC SYSTEM

A long time ago, before we were here,
A Frenchman named Mouton had an idea:
He said: we all have ten fingers as well as ten toes.
So, ten was the number that Mouton chose.
Ten is the number that will be the foundation
for a great measuring system for every nation.
There will be a good reason for metric popularity.
There are only three names to remember. Each one
has its own family.
We will show you how and why
Measuring with Metrics is as easy as pie!

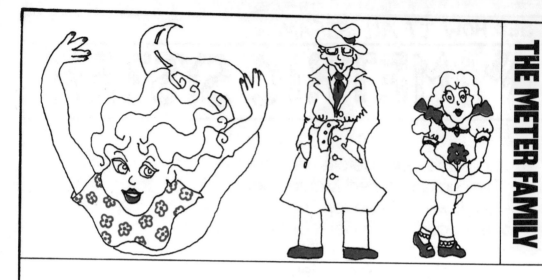

THE METER FAMILY

THE GRAM FAMILY

THE LITER FAMILY

What can you see that is as tiny as me?

You are bigger than a grain of salt, smaller than a sesame seed,
Much smaller than a peppercorn, you are very small indeed!

Exactly how tiny are you?

I am as tiny as a needle's eye,
as tiny as a freckle on your nose.
Or a little pencil dot
that hardly even shows.

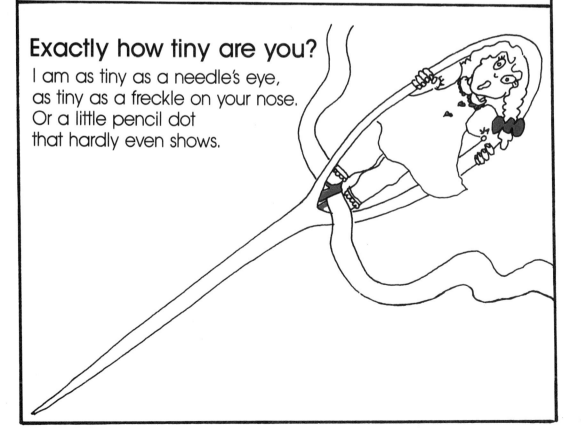

You seem too tiny to be important to me.
What are you used for? I would like to see.

I am used by electricians measuring a wire,
I am used by mechanics changing a tire.

I am used by jewelers setting a ring,
I am used by scientists examining a wing.

What can you see that is small like me?

You are bigger than a grain of rice, smaller than a cornflake.
Much smaller than the butter dish, well, for heavens sake!

Exactly how small are you?

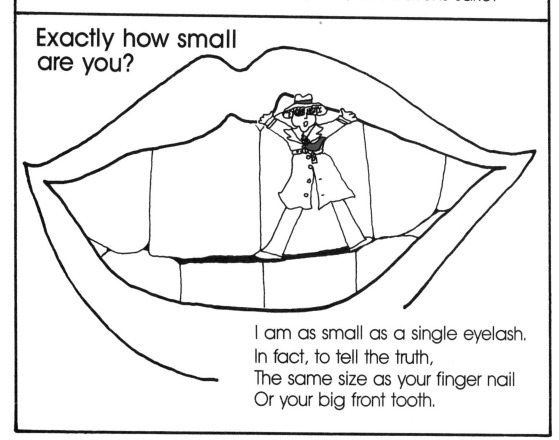

I am as small as a single eyelash.
In fact, to tell the truth,
The same size as your finger nail
Or your big front tooth.

You are bigger than a millimeter. I know that's true.
Who else uses you to measure? Give me a clue.

The centimeter is used by botanists who measure plants.
I am used by entymologists who measure ants.

Centimeters are used by architects to measure a small space.
I am used by dress makers when measuring lace.

I would like to introduce myself. I think I should confess
I am a metric marvel too, in case you did not guess.

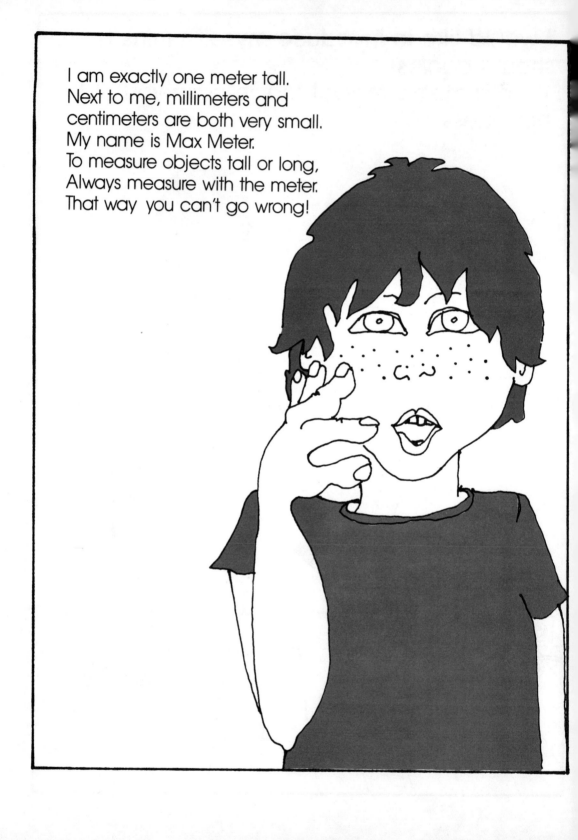

I am exactly one meter tall.
Next to me, millimeters and
centimeters are both very small.
My name is Max Meter.
To measure objects tall or long,
Always measure with the meter.
That way you can't go wrong!

What can you see as long or as tall as me?

I am taller than my baby brother, not nearly as long as the rug.
About the height of the piano, or the size of a great big hug!

Exactly how tall am I?

I come up to the top of Dad's desk.
I can see into the drawer.
It is just about one meter
From that door knob to the floor!

What measures one meter besides little old me? Stick around and you will see!

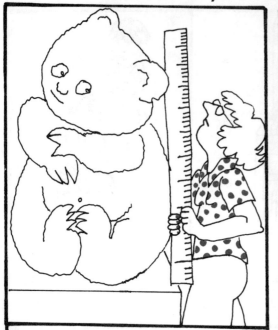

How else would zoologists measure bears?
How else would furniture makers measure chairs?

How else would pediatricians measure kids?
How else would archaeologists measure pyramids?

I was clean and shiny and feeling all right.
When a fighter walked in looking for a fight.

What can you see that weighs as much as me?

You are heavier than my rubber duckie, lighter than the bathroom scale.
Much lighter than my little ladder, or the diaper pail.

Exactly how heavy are you?

I weigh as much as a box of diapers.
That has not been opened yet,
As much as the first aid kit,
Or the plant on the cabinet.

I weigh heavy things, you will see without doubt.
I am used in so many ways, it will knock you out.

Kilograms are needed to weigh the farmer's crops.
Kilograms weigh the butcher's lamb chops.

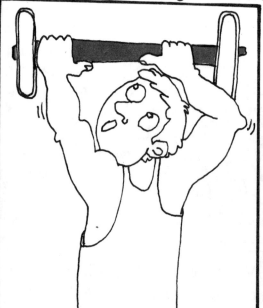

Weight lifters think kilograms when lifting a bar bell.
Sculptors think kilograms when purchasing their marble.

I stretch farther than you can see!
It takes 1000 meters to
make one of me!
I am Katie Kilometer.
When you go someplace
on foot or by car,
I can show you just how far.

Can you guess how far I go
From my head to my toe?

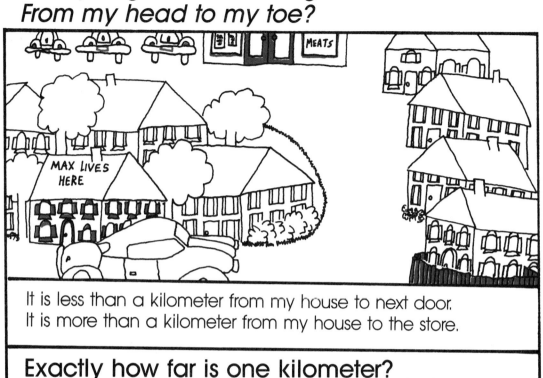

It is less than a kilometer from my house to next door.
It is more than a kilometer from my house to the store.

Exactly how far is one kilometer?

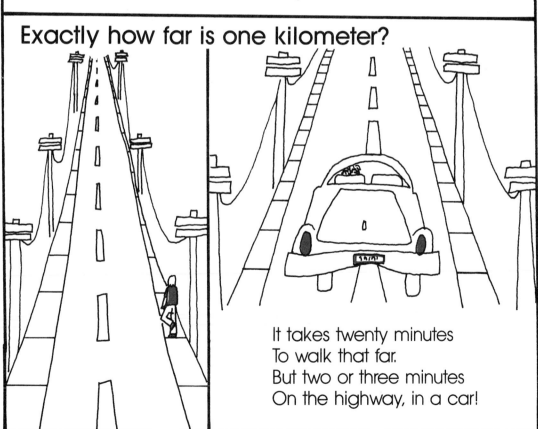

It takes twenty minutes
To walk that far.
But two or three minutes
On the highway, in a car!

*You have told me how far you go.
Now tell me why we need to know.*

A weatherperson needs me to show the wind's force.
A pilot needs me to chart his course.

An astronomer needs me to find out how far
It is from the earth to a distant star.

I am as light as can be,
You can hardly feel me.
I am Marty Milligram.
I am just right
For weighing something very light.

What can you see
That weighs as little as me?

You weigh more than a bubble of soap, less than a cotton ball,
Or a grain of baby powder. Your weight is very, very small.

Exactly how light are you?

I weigh as little as a bristle from your brush,
Or a strand of your hair,
Or an itsy-bitsy insect flying in the air.

What makes you think you are so terrific? How are you used? Please be specific!

I am used by pharmacists dispensing medication.
I am used by criminologists in an investigation.

I am used by a geologist assessing a stone.
I am used by an anthropologist analyzing a bone.

What can you see
That weighs one gram, like me?

You are heavier than a contact lens, lighter than Mom's barrette,
Much lighter than a bar of soap, or the washcloth when it's wet.

Exactly how much does one gram weigh?

I weigh as much as a hair pin,
Or as much as one bandaid.
I weigh as much as the rubber band
Mom uses for her braid.

Your examples are fine and dandy.
But tell me how you come in handy.

I am used by postal workers weighing mail.
I am used by the grocer weighing stringbeans or kale.

I am used by zookeepers weighing snakes.
I am used by bakers weighing cakes.

I am much more than a drop
That goes ker-plop!
I am Mitzi Milliliter.
Are you ready? Are you set?
To measure very small amounts
Either dry or wet?

What can you see with the same volume as me?

Are you more than the smallest dash of chili powder?
Are you less than a spoonful of tea? Much less than a cup of chowder.

Exactly how much is one milliliter?

I have the same volume
As five drops or one drip,
As much paint as this brush holds,
As much as one sip.

Would you further describe what you are about?
What are you used for day in and day out?

I am used by a nurse giving a shot.
I am used by a parent when feeding a tot.

I am used by a manicurist polishing nails.
I am used by a gourmet preparing snails.

What can you see
With the same volume as me?

Is a liter more than a large glass of milk, less than a kettle of stew?
Much less than a jug of cider, much less than a bucketful too?

Exactly how much is one liter?

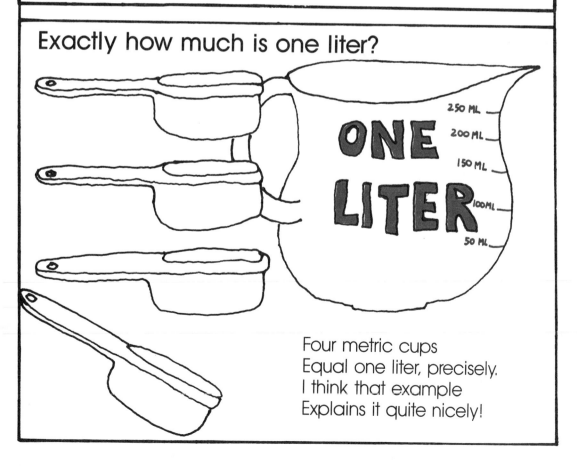

Four metric cups
Equal one liter, precisely.
I think that example
Explains it quite nicely!

Could you tell me how, why, and when the liter is used to measure again and again?

I am used by gas station attendants pumping gas,
By school teachers reviewing metrics in class.

I am used by vintners bottling wine,
By house painters pouring turpentine.

MAX'S METRIC MUNCHIES

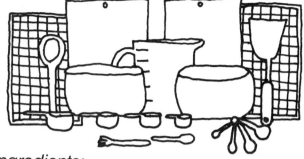

Ingredients:

- 120 grams butter
- 250 grams molasses
- 100 grams sugar
- 1 egg
- 300 grams flour
- 25 grams oatmeal
- 1 gram salt
- 4 grams baking soda
- 2 grams ginger
- 2 grams cinnamon

1. Beat butter with molasses, sugar, and egg in a large bowl with a wooden spoon, or in a mixer at medium speed, until the mixture is light and smooth.
2. Measure out the flour and pour into the other bowl. Add remaining ingredients to the bowl with the flour. Mix with a fork until blended.
3. Add the flour mixture to the molasses mixture and stir until smooth and well blended.
4. Lightly grease 2 or 3 baking sheets.
5. With a teaspoon, scoop up a little dough and push it off the spoon onto the baking sheets in a little mound. Press each lightly with the back of the fork to flatten.
6. Set the oven at 163°C (325°F) and let it heat up for 5 minutes. Place the cookies in the oven and bake for 12 minutes. Remove baking sheets from the oven and lift the cookies with a spatula onto the wire racks to cool. Makes 50 cookies.

Recipe by Sylvia Schur

Metric Conversion Chart

To calculate size—whether large or small, use this chart. It has it all!

Symbol	When you know	Multiply By	To Find	Symbol
mm	millimeters	0.04	inches	in
cm	centimeters	0.4	inches	in
m	meters	3.3	feet	ft
m	meters	1.1	yards	yd
km	kilometers	0.6	miles	mi

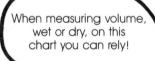

When measuring volume, wet or dry, on this chart you can rely!

Symbol	When you know	Multiply By	To Find	Symbol
ml	milliliters	0.03	fluid ounces	fl oz
l	liters	2.1	pints	pt
l	liters	1.06	quarts	qt
l	liters	0.26	gallons	gal